Atif Khan
Hassan Javed

Bleaching of Vegetable oil using Organic Acid Activated Fuller's Earth

AF167387

Atif Khan
Hassan Javed

Bleaching of Vegetable oil using Organic Acid Activated Fuller's Earth

LAP LAMBERT Academic Publishing

Impressum / Imprint

Bibliografische Information der Deutschen Nationalbibliothek: Die Deutsche Nationalbibliothek verzeichnet diese Publikation in der Deutschen Nationalbibliografie; detaillierte bibliografische Daten sind im Internet über http://dnb.d-nb.de abrufbar.

Alle in diesem Buch genannten Marken und Produktnamen unterliegen warenzeichen-, marken- oder patentrechtlichem Schutz bzw. sind Warenzeichen oder eingetragene Warenzeichen der jeweiligen Inhaber. Die Wiedergabe von Marken, Produktnamen, Gebrauchsnamen, Handelsnamen, Warenbezeichnungen u.s.w. in diesem Werk berechtigt auch ohne besondere Kennzeichnung nicht zu der Annahme, dass solche Namen im Sinne der Warenzeichen- und Markenschutzgesetzgebung als frei zu betrachten wären und daher von jedermann benutzt werden dürften.

Bibliographic information published by the Deutsche Nationalbibliothek: The Deutsche Nationalbibliothek lists this publication in the Deutsche Nationalbibliografie; detailed bibliographic data are available in the Internet at http://dnb.d-nb.de.

Any brand names and product names mentioned in this book are subject to trademark, brand or patent protection and are trademarks or registered trademarks of their respective holders. The use of brand names, product names, common names, trade names, product descriptions etc. even without a particular marking in this work is in no way to be construed to mean that such names may be regarded as unrestricted in respect of trademark and brand protection legislation and could thus be used by anyone.

Coverbild / Cover image: www.ingimage.com

Verlag / Publisher:
LAP LAMBERT Academic Publishing
ist ein Imprint der / is a trademark of
OmniScriptum GmbH & Co. KG
Heinrich-Böcking-Str. 6-8, 66121 Saarbrücken, Deutschland / Germany
Email: info@lap-publishing.com

Herstellung: siehe letzte Seite /
Printed at: see last page
ISBN: 978-3-659-77326-6

Dedicated to

My Parents, My Supervisor and My Lovely Brother Mr. Nasir Jamal

ACKNOWLEDGEMENT

All commend to Almighty Allah, the only real creator and owner of everything. All respects are for His Holy Prophet (PBUH), whose teachings are true source of acquaintance & guidance for whole mankind.

First of all I would like to thank Allah who has given me the courage and strength throughout my research work and helped me at all moments of my life. I am very thankful to my family for motivating me and for being a source of ethical support to complete my project.

I am very pleased to thank my supervisor who has been very helpful throughout my project and his suggestion and friendly criticism and motivation played a very important role for the completion of my project. He always supported me and contributed to bring best out of me.

I would also like to thank Managing Director Mr. Brij Kumar at Good Earth private limited for his dedicated help for arranging data and analysis for the project. I also take the opportunity to thank Mr Umair Saeed Production Manager at Priemere Dairy plant in Sheikhupura for his friendly contributions and guidance.

At University of Engineering & Technology, I would like to express my sincere gratitude to Prof Dr. Nadeem Feroze, Dean and Prof. Dr. Hassan Javed Naqvi, Chairman of the department. I am also thankful to my friends specially Mr. Zohaib Ashraf for moral support to complete this project.

And Finally, I would like to mention the names to whom I am really indebted; Prof. Dr. Hassan Javed Naqvi, my Supervisor for providing me technical guidance throughout the research and my parents for providing me motivation and inspiration. I understand without their friendly support it would not have been possible to complete this project.

Atif Khan

ABSTRACT

Specific properties of bentonite clay have made them a valuable material in different process industries. Easy availability, low price and their effectiveness are the major factors which made fuller's earth, an adsorbent in cooking oil manufacturing industry. Textural characteristics of fuller's earth play an important role in its performance. These characteristics can be modified by treatment with organic acids, which is the safest method for enhancing the properties of fuller's earth. Major motivation of this research is to increase the adsorption capacity of fuller's earth by modification in its properties by organic acid treatment. Organic acid treatment assures the safety of equipment and safety of the labor as well. Four organic acids are used which are recommended best for activation. These acids are acetic acid, phosphoric acid, citric acid and oxalic acid. Clay is treated with all of these four acids and the acid which generates more active sites in clay is the best recommended acid for activation.

Fourier transform Infrared spectroscopy (FT-IR) technique is performed for structural changes in clay structure, its crystalline properties and the expansion of basal spacing after the reaction with organic acid solution. Infrared spectroscopy is preferred over other testing techniques because it provides the rich array of absorption band which provide the wealth of structural information of clay sample. It covers the electromagnetic range of 500 cm^{-1} to 4000 cm^{-1}. FT-IR analysis may simply involve the characterization of material with respect to the presence or absence of specific group frequency associated with one or more fundamental modes of vibration. The spectral data is also used to measure one or more complex compounds in a simple or complex structure.

Table of Contents

List of Figures

List of Tables

Chapter 1

Introduction

---

1 Introduction

Due to unique properties of clay and clay minerals their value become more and more in wide range of industrial applications. Low cost, local availability and effectiveness are the key factors that have made clay and clay minerals to be used extensively as adsorbent for purification of vegetable oils. The physical properties of fuller's earth play an important role for the performance of bleaching earth. If these physical characteristics are modified through some techniques then the performance of these clays can be enhanced. These adsorbent clays which are used in process industries are known as bentonite clays. Bentonite is a clay mineral consisting essentially of montmorillonite which is mostly categorized into sodium (Na) or calcium (Ca) types depending upon exchangeable ion. The adsorptive properties of bentonite depend upon high surface area, swelling capacity and cat-ion exchange capacity. These properties are related to crystal isomorphic substitution [1].

1.1 Vegetable oil processing

1.1.1 Refining

The usual process of cooking oil consists of degumming, bleaching, hydrogenation and deodorization. These steps are carried out in batch operations, except of a few are carried out in continuous processes operation. The oils are degummed by some coagulants with a small amount

(0.5%) of concentrated phosphoric acid. If alkali refining methods are used, the free fatty acids are neutralized with an excess of 0.1% sodium hydroxide solution at 75^0C to break the emulsions if produced. The gums and soaps are removed by centrifugal action process, and the fatty acids are recovered by acidification. If steam refining is involved then degumming is carried out along with bleaching if necessary, and treatment with sparging steam at high vacuum conditions so that the fatty acids are removed by distillation process.

1.1.2 Bleaching

Bleaching is done by the use of adsorption process through bentonite clays for vegetable oils, and alternative method by chemical reactions for nonedible oils. The oil so bleached is then subjected to refrigeration treatment which removes impurities through solidification at refrigerator temperatures. This is done by cooling at 5^0C and filtering out any solidified material.

1.1.3 Hydrogenation

Hydrogenation or solidification, applied to fats and oils, is defined as the conversion of various unsaturated glycerides into more highly completely saturated glycerides by the addition of hydrogen using the catalyst. Various types of oils, such as soybean, cottonseed, fish, whale and peanut are converted through partial hydrogenation into fats of a composition of more suitable for animal shortenings, butter margarines, and other edible purposes, as well as for soap manufacturing and for variety of other industrial uses. The objective of hydrogenation is not only to increase the melting point but also to improve the qualities, taste and odor for different types of edible oils. It is frequently accompanied by process of isomerization with tremendous increase in melting point, caused, for example, by oleic (cis) isomerizing to elaidic (trans) acid. As the reaction involves is exothermic, the major amount of energy required for the production of hydrogen, warming of the oil, pumping and filtering. The reaction may be generalized.

$$(C_{17}H_{31}COO)_3. C_3H_5 + 3H_2 \longrightarrow (C_{17}H_{33}COO)_3.C_3H_5 \quad \Delta H = \text{-420.8 KJ/kg}$$

Manufacture of hydrogenated oils requires the equipment in which hydrogenation is carried out, catalyst, equipment for refining the oil before hydrogenation, a converter in which actual hydrogenation is carried out, and equipment which is required after hydrogenation for treatment of the fat. The hydrogen manufactured by different methods, but the process which is used most widely is the stemi- carbon process for hydrogen. Since sulfur compounds (H_2S, SO_2 etc.) are major source of catalyst poisoning even in very small amounts, as is carbon monoxide, but to a lesser degree, it is essential that the hydrogen should be completely free of these poisons, as well as the taste producing materials. The amount of hydrogen necessary is directly proportional to the degree of reduction in unsaturation required. The catalyst and unsaturated oil are introduced into a specially designed reactor. The charge is heated as quickly as possible to as high as 240^0C, but 190^0C is a more common temperature. Normal operating pressures are 200 to 700 KPa gauge. At about 150^0C the nickel formate begins the reduction.

$$Ni\,(HCOO)_2 . 2H_2O \longrightarrow Ni + 2CO_2 + H_2 + 2H_2O$$

The charge is maintained at high temperature for almost 1hour and then cooled. During the reduction and cooling time, the hydrogen is bubbled through the oil slowly to remove decomposition products from the oil. At completion of the reduction reaction, the charge is pumped to the converter or formed into blocks, flakes, or granules for further use.

The hydrogenation is followed and controlled by refractometer readings to indicate the physical properties (saturation and melting point). The catalyst is filtered and reused. As in the hydrogenation reaction heat is evolved so the heat should be removed by heat exchanger. Selective hydrogenation can also be used, where polyunsaturated fatty acids can be largely converted to mono-saturated acids before there is significant conversion of the mono-unsaturated fatty acids to saturated fatty acids. Also, conditions can be changed to permit hydrogenation of both mono and poly- unsaturated fatty acids at the same time.

1.1.4 Deodorization

Deodorization process is carried out by using the blow of superheated steam through the oil (if hydrogenated, and it is l hot and in the liquid phase) under a high vacuum of 138 to 800 Pa and 210 to 275^OC. This removes most of the compounds which causes odor and also finishes most of the color producing pigments. Final packaging is often carried out under a nitrogen atmosphere to prevent any deleterious oxidation [2].

1.2 Scope of Work

The endeavor of this project was to develop the safest technique for activation of fuller's earth and creating the best activated clay without any hazards to both the equipment and the labor. In this way the process of cooking oil purification will be least hazardous and give the maximum confirmation of labor and equipment security throughout the process. Another objective of this project is to purify the used cooking oil with the same organic acid activated clay for analyzing the maximum adsorbing capacity of clay to get maximum efficient and highly activated organo-clay.

Scope of work includes:

- ✓ Clay Sample preparations
- ✓ Organic acid treatment of these samples
- ✓ Characterizing these activated clay samples
- ✓ Selection of the clay which shows maximum adsorbing capacity

1.3 Structure of Thesis

The thesis is divided into following sections.

Chapter 1 gives the introduction and overview of fuller earth and its importance in cooking oil industry. This chapter also covers the major processing steps in refining and the portion of the process where fuller's earth is used is also highlighted. Chapter 2 covers the literature review, means how much research is carried out till now on activation of fuller's earth. 2nd chapter will also cover historical background of fuller's earth its discovery in ancient times and how it was used in different applications. Chapter 3 will discuss the methodology of the research; means which steps are carried out for making the activated fuller's earth also major processes which involve in creation of activated fuller's earth will also be discussed. In chapter 4 analysis of the activated fuller's earth will be discussed. Techniques which are used for analysis will be discussed. Finally in chapter 5 conclusion and recommendations will be covered.

Chapter 2

Literature Review

---

2 Literature Review

2.1 Historical Background:

The English word "fuller" derives from the Latin term "fullo", showing that the person whose job was to degrease and thickening the clothes. Prior to the discovery of fuller's earth two detergents were occasionally discovered i.e. the river mud and old urine. The first one must be found by washing (sheep's) fleece in the river. The second one i.e. old urine was perhaps discovered by a woman who observed that the baby's bed which got wet became whiter if washed after a couple of days, especially that relevant part which became wet with the urine.

The woolen cloth washing with fuller's earth was mostly done in Cyprus about 5000 B.C. Sheep's were completely domesticated for almost thousand years before. They were found in houses constructed of lumps of fuller's earth, in 1969 carbon-dated about 5000 BC. The particular greenish grey clay was rich in montmorillonite.

In Mesopotamia the fuller's earth was not of very good quality. It might be possible at that time; fleeces and clot were filled in river mud in ancient times. Other detergent materials were being used later on. Detergents materials such as wood ashes or soda were used when washing in the la tub using fuller earth stirred with the contents having a staff pole because the alkali damaged the feet skin. Finally a special bath (mazuru) was used to beat the cloth.

Indications of the use of fuller's earth, old urine as detergent and cloth washing can be found out in Bible, particularly in the Old Testament. In the 20th century, In east of Jerico, a mix of clay and ash was used if we go back to 3000 BC. In this settlement a local people used to wash cloths at that time, the name of the place has an Arabian verb for to "wash". Moreover, within the same area a clay source was not still found. Even today, Israel does not have its own clay reserves and it still imports clays from Cyprus [3].

2.1.1 Bleaching Earths

An American tourist came in the Middle East and observed that the people used clay to lighten edible oils. Fairnanks & Co. in Chicago tested the clays for bleaching cotton seed oil. The best results found with the English Fuller's earth. This earth was then used for bleaching vegetable oils in USA since 1878, and became valuable in the European and American Market. However, ten years later, John Olson found clays reserves in Arkansas (1891), which were identical to the English Fuller's earths. After few years large deposits were found in Quincy in Florida, the monopoly of the English fuller's earth were broken. The Florida's Earth (about 33000 tons in 1912) was mainly used in oil refining. Vegetable and animal oil were extracted with English fuller's earth (about 20,000 tons in 1912) before the advent of First World War.

Prior to bleaching with fuller's earths, all types of oil, fats and waxes were bleached with different substances, like, prussiate of potash and animal charcoal. These expensive materials absorbed the coloring components of oil but also adsorbed huge amounts of the oil itself (up to 2.5 gram oil per gram of adsorbent). Application and manufacturing of bleaching earths increased in USA, and in 1924 almost all material was explored and produced in South-States of USA. The manufactured fuller earth was divided into several fractions, but unfractionated material was also used.

As only minor deposits of fuller's earth were discovered in Germany (Saxonia, Westerwald, Silesia), so several thousand tons of American and English fuller's earths were imported. Since 1924 modified clays from DeathValley, bentonite from California, and Montana etc. began to threaten the markets of fuller's earths. Following an American patent these bentonites were treated with sulfuric acid and become superior to natural fuller's earths [3].

2.1.2 Organic activated bleaching earths

John W. Jordan Jr. (1912-2001) was the first scientist who created a research on organic activated bentonites (1949), these bentonites are known today as bentones. Organic activated bentonites are used in many applications.

Table 2-1 Application of Organic activated bleaching earth

Area of Applications	Uses
Chemical Process Industry	Insecticides and fungicides support
Metallurgical Foundries	Binding agents for anhydrous sands and for blackwashes thickening
Mineral oil processing	Decolorizing, refining and purification of drilling fluids and for grease thickening
Paints and Varnishes manufacturing	Thickening of varnishes, paints, coating materials, waxes and adhesives
Pharmaceutical and cosmetics industries	Odor control and liquid adsorption
Tar Exploitation	Tar and asphalt coatings

2.2 Clay structure and its properties

There are numerous literatures in which author point of views about clay and clay minerals, as well as their applications. It is compulsory to our current understanding that how and why clay minerals have industrial utilization on large scale. The terms clay and clay mineral are considered very different in their meaning in this context. Association Internationale pour l' Etude des Argiles (AIPEA) nomenclature committee defined clay as "naturally occurring constituent consist mainly of fine grained minerals which is generally plastic at relevant water contents and becomes harden when dried or fired". The definition of clay minerals according to the AIPEA refers to "silicates minerals which impart plasticity on clay which harden upon drying or firing". Clay and clay minerals are used for low-cost adsorptive materials due to its diversity in different areas, such as adsorptive materias in the bleaching of cooking oils, catalyst beds,medications, in the preparation of pillared clays, carbonless copy paper, textile industries, petroleum processing & refining, etc.

The molecular structure is the most important factor for different type of clay minerals. When one octahedral sheet is attached to the other octahedral sheet, the clay layer formed is known as 1:1 layer type, such as serpentine subgroup and kaoline subgroup. Within 1:1 layer type, there are various mineral types, for example di-octahedral (e.g kaolinite and halloysite) with a composition of $Al_2Si_2O_5$ (OH)$_4$ and tri-octahedral (e.g. Antigorite and chrysotile) with an ideal composition of $Mg_3Si_2O_5$ (OH)$_4$. The composition of Kaolinite and holloysite are classified by a predominance of Al^{+3} in octahedral sites, although some isomorphs substitution if Mg^{+2}, Fe^{+3}, Ti^{+4} and V^{+3} for Al^{+3} may occur. However for antigorite and chryosite, it is tri-octahedral layer minerals and it contains mainly Mg^{+2} as central atoms in the octahedral sites.

The structure formed when two octahedral sheets are sandwiched on octahedral sheet is called 2:1 layer type, such as smetite sub group, micas subgroup, pyrophyllite subgroup, vermiculite subgroup and chlorite subgroup. Among these clay minerals subgroup, the most commonly used in different chemical process industry is smectite subgroup. This is due to high specific surface area, high adsorptive capacity stability in chemical composition, high cation exchange capacity and swelling characteristics [4].

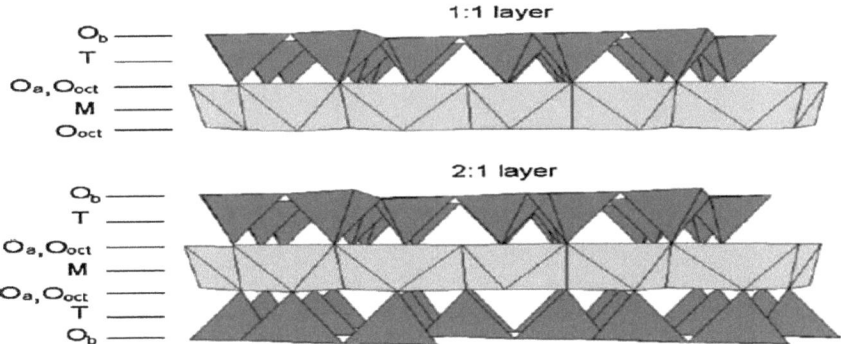

Figure 2-1: The clay mineral structure of 1:1 and 1:2 layer type, Where Oa, Ob and Ooct refer to tetra-hedral basal, tetra-hedral apical and octa-hedral anionic position respectively. M and T show the octahedral and tetra-hedral cation, respectively (adapted from Brigatti et-al) [5].

2.3   Structural Characteristics

The structural characteristics such as surface area and pore volume play major role in the bleaching earth performance. These above mentioned factors can be improved by making some modifications in the properties of clay minerals using various techniques including acidic, basic, organic and thermal treatment. Both chemical and physical characteristics of fuller's earth are dependent on its structure and molecular arrangement. For learning, the structure and molecular arrangement of Kaolins, Smectites and palygorskite-sepiolite are entirely not same even though the fundamental building blocks, i.e. the tetrahedral and octahedral sheets, are similar. Smectite and palygorskite are widely used as adsorbing agent in the purification of edible oils. Main reason is due to their very small size, high available area on surface for adsorption and medium capacity of ion exchnage which give smectite, palygorskite and sepiolite, a high capacity to adsorb different types of liquids. According to different researchers the specific surface area can be enhanced by acidic treatment compared to natural or untreated clay. The authors also reported that natural smectites have a specific surface area ranging from $16m^2/g$ to $97m^2/g$ and shows lower adsorption capacity to adsorb coloring matter and other impurities in oils and solutions. In addition to that, the surface area of smectites is several times larger in acid treated clay, reaching values $200-400m^2/g$. Compared to untreated clay, acid activated clay showed more adsorbing activity on its surface.

Although the bleaching power increases with increasing surface area, but it depends more on size distribution of the bleaching earth. The adsorption capacity of clay depends on the accessibility of the molecules into the pores and respectively on size of the pores. According to Gregg and Sing pore sizes can be categorized into three types. The pores of widths below 2nm are called micro-pores, those with widths between 2nm and 50nm are called meso-pores and those greater than 50nm are called macro-pores. Properties of bentonites such as adsorption and catalytic activity depend extensively on micro-pores and meso-pores. Although the micro-pores and meso-pores are located inside the particles and the macro-pores are located in between the particles, nevertheless the effect of macro-pores on the adsorption capacity of solid is almost at negligible level compared to that of micro-pores and meso-pores. The bentonites clay contains natural meso-pores and small amount of micropores [6].

2.4 Surface chemistry

For identification of the mechanism of adsorption of clay minerals, studies on clay surface (both outside and intermolecular) have been investigated by Schoonheydt and Jhonston, which help to clear the point of view about clay's surface, such as adsorb-ate, and its location on the adsorbent etc. Generally, clay minerals have many useful characteristics in its physique, namely; active sites such as hydroxyl groups, Lewis and Bronsted acidity, exchangeable cations and the difference of SiO_4 tetrahedral sheet and Al_2 (OH) $_6$, octahedral sheet in chemical stability. Studies of clay surfaces till now have considered active sites and the functional groups. Active sites and its surface functional groups are of great importance, because they determine the chemical processing. The active site is an atom or group of atoms that is relevant to the solid and can take part in reactions with the particles in the surrounding media. When the protonation & deprotonation reactions at the surface of silicates are considered, the focus is on the surface oxygen instead of ionic metals, because these are the atoms that can become protonated & de-protonated.

The chemical reactivity of surface oxygen depends on the type and the distribution of atoms surrounding oxygen; thus a functional group attached to a surface is usually written as single oxygen together with the atoms that are linked   to it. Therefore different planar surfaces have different functional groups. They can also be explained on the basis of their location (edge or basal surfaces), geometric configuration of surface atoms, chemical composition and access. Generally there are six kinds of sites which are useful in explaining the interaction of organic molecules to the surfaces of clay; they are isomorphic substitution sites, hydrophobic sites, broken edge sites/hydroxyl surface, neutral siloxane surface, metal cations occupying cation exchange sites and water molecules surrounding the exchangeable cations. [7].

2.5 Techniques of Activation

Activation is defined as a treatment applied to different types of clays to enhance a capacity for a decolorizing and removal of other impurities in oils and solution. Physical and chemical characteristics of clay minerals are studied by many researchers relevant to its relation to their adsorbing and catalytic properties. This study is governed by the extent and nature of their outer surface which can be changed by suitable treatments. The treatments or modification of clay minerals can be categorized in following diagram.[8].

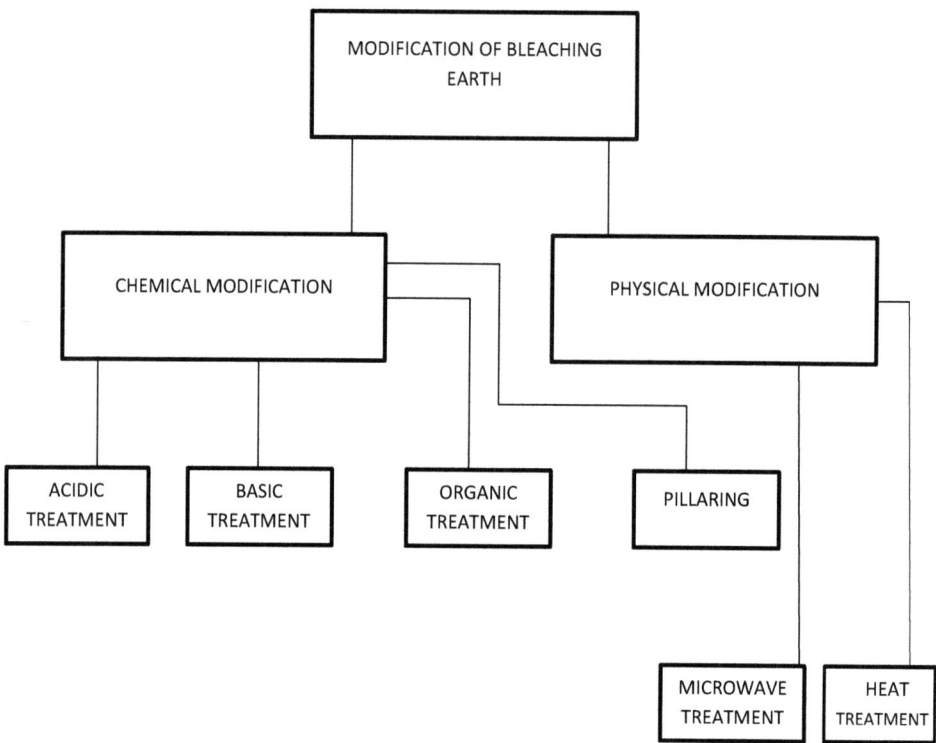

Figure 2-2: Categorization of activation treatment technique

2.6 Modification of chemical characteristics

Acid treatment is the most famous technique used for the modification in adsorptive and chemical properties of bleaching earths. This technique consists of two steps; (1) the substitution of exchangeable cations by protons and (2) the dissolution of metals ions from the clay structure, by depopulating the octahedral sheet (figure 2.3).

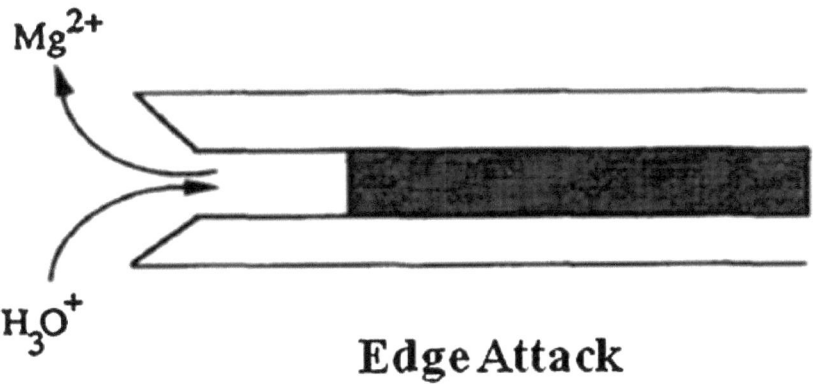

Edge Attack

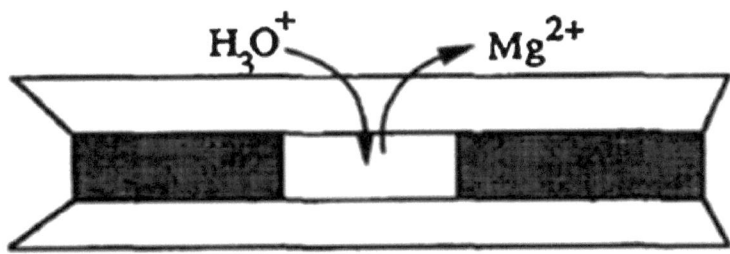

Substitution of exchangeable cations

Figure 2-3: Mechanism of acid activation

The bleaching capacity depends on the activation methodology steps (acid concentration, temperature, time). The clay composition (layer composition) affects the chemical stability against the acid attack; with tri-octahedral layers disintegrate faster than di-octahedral layers. Increased Mg and / or Fe for Al substitution in dioctahedral smectites, increases their dissolution rate in acids. The modification with acid is used to enhance the surface area, alterations in the surface functional group for formation of solids with maximum porosity and to improve the number of acidic centers. The most important role is to modify the adsorption capacity for the removal impurities and color pigment from the oil. Numerous studies have been reported on the acid treatment of clays. Among all the clays, bentonite is the most commonly used for the production of acid activated clays in the purification of vegetable oils. Bentonite has always a great potential of marketing and acid activated bentonite was the famous product for many decades.

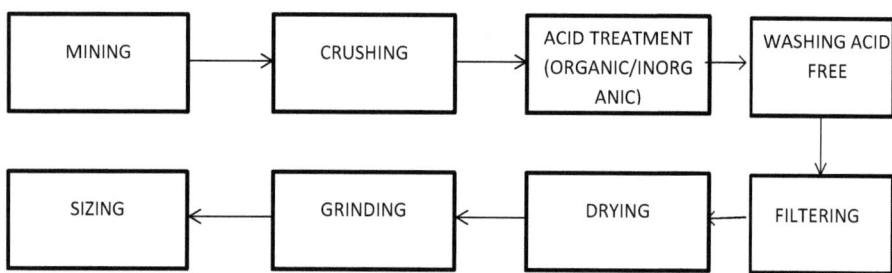

Figure 2-4: Simplified flow diagram of the process of acid activated bleaching earth

Treatment of clay with inorganic acids such as H_2SO_4, HCL generates new surface acid sites faster, but they also cause tremendous leaching of Al_2O_3 from the clay mineral, which ultimately leads to structural destruction. If organic acids are used instead or inorganic acids, it may preserve the clay structure, but are less effective in generating surface acid sites. Moreover organic acids dissolve less of Al_2O_3 and thus its activation power is less than the mineral acid but some of the organic acids are compatible to inorganic acids and can be handled in a safe way [9].

2.7 Modification of physical characteristics

Other than chemical alterations, the physical modifications using heat treatment also play a major role for modifying the structure of bleaching earth. The structure and molecular arrangement of clay can be altered by heating at high temperature. Some physicochemical properties of bentonites such as expansion due to swelling, strength, cation exchange capacity, particle size, specific surface area and surface acidic centers can be changed by thermal treatment [10].

2.8 Optimum Operating conditions

Although numerous researches has been done on activation of bleaching earth, there is still lack of information on a full study relating to the optimum operating parameters for obtaining the high capacity bleaching earth. Therefore, a terminology known as design of experiments (DOE) especially response surface methodology (RSM) is a powerful and effective tool for process analysis, optimization and modeling. This methodology involves the process planning and making a setup of experiment. The collected data is accordingly analyzed and interpreted. The approach used by this technique is to reduce the number of performed experiments, improves statistical interpretation and finding the optimum parameters for acid activation. As with most processing tools, many parameters affect the performance of bleaching earth. However, a detailed study on the DOE methodology cannot be discussed due to the lack of research on the variables using this technique. Nevertheless, this study takes into account the effects of parameters on the activation treatment of bleaching earth. There are many factors affecting the performance of bleaching earth which includes adsorbent type and dosage, effect of time and temperature on bleaching, effect of vacuum, effect of agitation, oil retention, effect of moisture and effect of particle size [11].

Chapter 3

Methodology

3 Methodology

The research work has been done in the following steps

- ✓ Process sturdy
- ✓ Sample preparation
- ✓ Sample analysis
- ✓ Results and its comparison
- ✓ Final Conclusion

3.1 Process Study

There are two major processes study required for this research. (1) Vegetable oil processing and its manufacturing steps which is discussed with detail already in chapter 1. (2) Activation process of fuller earth. Activation of fuller earth through organic acid is also discussed in detail through strong literature review in chapter 2. Knowing the process detail is very important to know the importance of fuller's earth, its application and the method of its utilization during vegetable oil processing.

3.2 Sample preparation

The fuller's earth for activation and analysis was taken from the "Good Earth" company in Sheikhupura (Pakistan) which manufactures the activated fuller's earth or bleaching earth for different vegetable oil processing industries. 100 grams of weighed fuller's earth are taken separately for sample preparation. Each of these samples was ground to powder form. Then these samples were treated with four organic acids which are most suitable for activation according to the literature. These four organic acids were oxalic acid, phosphoric acid, citric acid, acetic acid. These acids with different concentrations are used but 1N acid solution is optimum for treatment and showed satisfactory results so 1N acid solution is use for treatment.

Figure 3-1: Major steps involved in Research Methodology

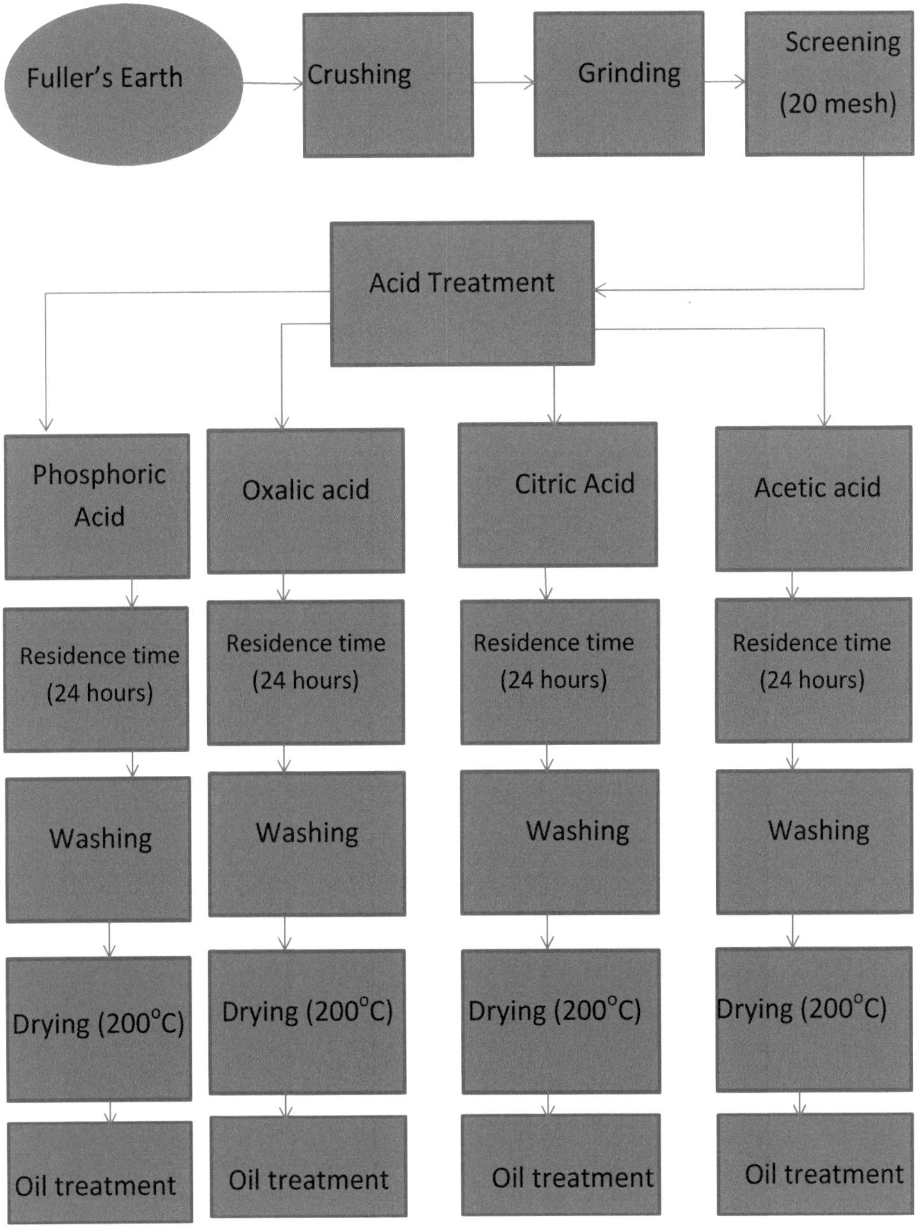

3.2.1 Phosphoric acid treatment

Phosphoric acid is best recommended for the removal of aluminum ions. It reacts with aluminum and aluminum phosphate is formed, which settles down and easily removed. Literature shows that treatment of clay in Algeria with 70% acid dosage and 24 hours residence time, 66 % conversion of bleaching earth is achieved. Concentration of acid and residence time both highly depends upon the nature of the clay. When maximum conversion is achieved then bleaching earth is washed with water so that acid is completely removed. Then in next step it is dried in 200C for almost complete removal of moisture. Then its adsorption capacity is tested by treating with edible oil.

3.2.2 Oxalic Acid treatment

Second sample of fuller earth is then treated with oxalic acid. According to literature survey oxalic acid is good for removal of both aluminum and iron. Research on Brazilian clay reveals that 81% oxalic acid dosage and 24 hours residence time gives 74% conversion. In this research different dosages of oxalic acid are selected to obtain maximum conversion. After acid treatment it is washed with water and dried at 200C for maximum removal of moisture. Finally adsorption capacity is tested by treating it with edible or vegetable oil.

3.2.3 Citric Acid treatment

Citric acid bleaching is mostly recommended when soybean oil refining is required. Literature does not reveal conversion of bleaching earth using citric acid. Therefore different dosages of acid will be checked for maximum conversion of bleaching earth. When maximum conversion is achieved then same procedure of washing and drying is adopted as mentioned above and then finally its adsorption capacity is tested by oil treatment.

3.2.4 Acetic Acid treatment

Acetic acid is recommended when palm oil refining is required. According to literature 1N acetic acid with 0.5 hours' time give 66% conversion. Acetic acid is recommended for removal of magnesium. By varying its concentration and residence time conversion rate also changes. So that concentration will be selected at which maximum conversion occurs. After that same procedure of washing, drying and oil treatment is adopted for testing adsorption capacity.

As compared to inorganic acids these organic acids are expensive and have less adsorption capacity but safety of process and manpower is not compromised if organic acids are used for bleaching step. Moreover if some traces of these organic acids found in edible oil after refining then it is not as harmful for human body as inorganic acid traces. Finally the acid which has maximum bleaching capacity and less expensive will be recommended for bleaching [13].

3.3 Sample analysis

After preparation of clay and oil samples these samples were analyzed using most advanced technology for good and accurate results. This technology and sample analysis will be discussed in detail in next chapter.

3.4 Utilization of research result

Both organic and inorganic acids are used for generating active sites or surface activation of fuller earth. These results are compared and the organic acid which is most compatible to inorganic acid for activation of fuller earth will be the recommended acid for activation. The best recommended organic acid will be concluded in the next chapter.

Chapter 4

Results and Discussion

4 Results and discussion

After preparation of clay and oil samples, analysis of these samples had been carried. The study on these samples gave the clear view of untreated as well as treated clay. Analysis of the sample is performed with the following sequence.

- ✓ Analysis of untreated/raw clay
- ✓ Analysis of clay after activation treatment with acetic acid
- ✓ Analysis of clay after activation treatment with citric acid
- ✓ Analysis of clay after activation treatment with phosphoric acid
- ✓ Analysis of clay after activation treatment with oxalic acid
- ✓ Conclusion and recommendations

4.1 Method of Characterization

The analysis technique which is used for the characterization of both the untreated/treated clay is the Fourier's Transform Infrared Spectroscopy (FT-IR). The quality features of infrared spectroscopy are one of the most effective tools of this vast and advanced method for characterization. For so many years, large research has been done in terms of the basic frequencies for absorption (also known as group frequencies) which is very important tool for understanding of the structure and spectral co-relation of the associated molecular vibrations. Application of this precious information at serves to be a combination of both art and science.

4.1.1 Origins of the Infrared Spectrum

If the conventional terminology is used, the infrared spectrum is obtained as a result of the absorption of electromagnetic radiation at frequencies that are related to the vibrational motion of various chemical bonds from within a molecule. It is very important to understand the distribution of energy possessed by a molecule at any given moment. Total energy is defined as the sum of the contributing energy terms (Equation 4.1.1)

$$E_{total} = E_{vibrational} + E_{rotational} + E_{translational} + E_{electronic} \ldots\ldots\ldots (4.4.1)$$

In above equation the translational energy associated with the movement of molecules in space is dependent of the normal thermal motions of matter. Rotational energy, associated to its own spectroscopy form, is defined as the tumbling motion of a molecule, which is the consequent of the absorption of energy within the microwave region. The vibrational energy is a high value energy relation and related to energy absorption by a molecule as the atomic component that vibrates about the mean center of their chemical bonds. This particular electronic part is linked to the energy variations of electrons as they are spread on the molecule, either localized within particular bonding, or structurally delocalized, such as an aromatic ring. For analyzing such electronic transitions, it is important that energy application should be in the form of visible and ultraviolet radiation (Equation 4.4.2)

$$E = hv \; (frequency/energy) \ldots\ldots.. (4.4.2)$$

The basic need for infrared spectroscopy is leading to infrared radiation absorption, is that there must be a net change in dipole interaction during the vibration for the molecule or the functional group under observation.

4.1.2 Spectral Interpretation by Application of Vibrational Group frequencies

In this section data relative to the most specific functional group frequencies are presented in the form of tables. Most common functional groups are found in organic compounds .References are also used for basic inorganic compounds, in ionic species form.The purpose of this tabulated data is only a part of the interpretation process, and other aspects of the spectrum must be considered. For the sake of understanding of infrared spectroscopy, it is necessary to start from the basics of organic compounds, namely the fundamental backbone or the original hydrocarbon structure. Let us start with the simple, aliphatic hydrocarbon, which is at the simplest of most straight chain compounds. Straight chain hydrocarbons exist in simple linear chains, branched chains and in cyclic structures. Molecule may exist with one or more of these component structures. The infrared spectroscopy may provide information about the presence of these structures.

The amount of unsaturation in the form of a multiple bonding has a deep impact on the chemistry of the molecule, and likewise it has a strong effect on the spectrum so formed. Similarly, the same process occurred when a ring structure is present within a molecule. Infrared spectroscopy is an effective tool for identification of the presence of these functional groups. It gives the relevant information specific to the particular functional group itself, and also on the interrelation of the group with other sections of the molecule and on the spatial location of the group. It includes relation between a double bond and another unsaturated center, a cyclic ring or a group, such as a carbonyl (C=O), and the arrangement or location of the double bond within the molecule, such as *cis* or *trans* bonding. It should be noted that *cis*/*trans* relations are not relevant only to unsaturated hydrocarbons, and the terminology is applied anywhere else, such as with secondary amide structures. Again, the interrelated changes in the spatial distribution of the groups associated are reflected in the infrared spectrum as additional bands.

As we go further to simple organic compounds, where more than one functional group is combined to the molecule, many changes are observed in the spectrum. These results from the bonding relevant to the specific functional group, and also local involvement to the spectrum that links again to spatial variations and also to local and neighboring electronic configurations. Examples of such functional groups are halogens, simple oxygen type, such as hydroxyl and ether groups, and amino compounds. Carbonyl compounds, where the functional group addition indicates the C=O bond, also provide very basic contributions to the spectrum, and because of the diversified nature of these compounds they are best categorized in a separate group.

A very specific group of compounds, from a spectral point of view, are the double or triple bonded nitrogen compounds, such as cyanides and cyanates. These have very specific absorption frequencies, which are easy to relate, and are free from spectral disturbances. The same process can be observed for some of the hydrides of heteroatoms, such as sulfides (thiols), silanes, and phosphines. Finally, there are some other functional groups containing oxygen, as encountered in the nitrogen-oxy (NOx), phosphorus-oxy (POx), silicon-oxy (SiOx), and sulfur-oxy (SOx) compounds. These are sometimes not easy to identify from first principles, and a knowledge of the presence of the heteroatom is helpful. The spectra so obtained are relevant, but many of the oxy absorptions occur within a crowded and highly overlapped region of the spectrum, mainly between 1350 and 950 cm^{-1}. Also, majority of these compounds exibits C-O bonding, which is common in other frequent functional groups such as ethers and esters [14].

Following tables [14] shows the detail of some functional groups against their characteristic wave number.

Table 4-1 Aliphatic hydrocarbons functional groups

Wave Number	Functional group relevant to wave number
2860-2970	Methyl group C-H stretch
1370-1470	Methyl group C-H bend
1365-1385	Di-methyl or iso-doublet
1385-1395	Tri-methyl multiplet
2845-2935	Methylene C-H stretch
1445-1485	Methylene C-H bend
925-1055	Cyclo-hexane ring vibrations
700-1300	Skeletal C-C vibrations
2815-2850	Methoxy methyl ether O-CH$_3$, C-H stretch
2780-2820	Methyl amino, N-CH$_3$, C-H stretch

Table 4-2 Olefinic (Alkene) functional groups

Wave Number	Functional group relevant to wave number
1620-1680	Alkenyl C=C stretch
1625	Aryl substituted C=C
1600	Conjugated C=C
3075-3095	Vinylidene C-H stretch
3010-3040	Medial Cis or Trans C-H stretch
1410-1420	Vinyl C-H in plane bend
1290-1310	Vinyldene in C-H in plane bend
890-995	Vinyl C-H out of plane bend
885-895	Vinylidene C-H out of plane bend
960-970	Trans C-H out of plane bend
700 (broad)	Cis C-H out of plane bend
3010-3040	Terminal Vinyl C-H stretch

Table 4-3 Aromatic Ring (aryl) functional group

Wave Number	Functional group relevant to wave number
1580-1615	Aromatic ring stretch
1450-1510	Aromatic ring stretch
3070-3130	Aromatic C-H stretch
950-1225	Aromatic C-H in plane bend
670-900	Aromatic C-H out of plane bend
690-730	Mono-substitution (phenyl)
735-770	1,2-Disubstitution (ortho)
750-810	1,3-Disubstituiton (meta)
800-860	1,4-Disubstitution (para)
1660-2000	Aromatic combination bands

Table 4-4 Acetylenic (Alkyne) functional group

Wavenumber	Functional group associated to wavenumber
2100-2140	Terminal Alkyne (mono-substituted)
2190-2260	Medial Alkyne (di-substituted)
3310-3220	Alkyne C-H stretch
610-680	Alkyne C-H bend
630	Alkyne C-H bend

Table 4-5 Aliphatic organo-halogen functional group

Wavenumber	Functional group associated to wavenumber
1000-1150	Aliphatic Fluoro Compounds C-F Stretch
700-800	Aliphatic Chloro Compounds C-Cl Stretch
600-700	Aliphatic Bromo Compounds C-Br Stretch
500-600	Aliphatic Iodo Compounds C-I Stretch

Table 4-6 Alcohol and Hydroxy fictional group

Wavenumber	Functional group associated to wavenumber
3200-3570 (broad)	Hydroxyl group H- bonded OH stretch
3200-3400	Normal Polymeric OH stretch
3450-3550	Dimeric OH stretch
3540-3570	Internally bonded OH stretch
3600-3645(narrow)	Non bonded hydroxyl group OH stretch
3630-3645	Primary Alcohol OH stretch
3620-3635	Secondary Alcohol OH stretch
3540-3620	Tertiary Alcohol OH stretch
3530-3640	Phenols OH stretch
1310-1410	Phenol or Tertiary Alcohol ,OH bend

Table 4-7 Ether and oxy compound functional group

Wavenumber	Functional group associated to wavenumber
2810-2820	Methoxy, C-H stretch (CH₃-O-)
1050-1150	Alkyl substituted ether, C-O stretch
1070-1140	Cyclic ethers, large rings, C-O stretch
1230-1270	Aromatic ethers, aryl-O stretch
820-890	Peroxides, C-O-O- stretch

Table 4-8 Amine and amino compound functional group

Wavenumber	Functional group associated to wavenumber
3380-3400	Aliphatic primary amine
3460-3510	Aromatic primary amine
1590-1650	Primary amine NH bend
1020-1090	Primary amine CN stretch
3310-3360	Aliphatic secondary amine
3430-3490	Hetro-cyclic amine
3320-3350	Amino compounds NH stretch
1550-1650	Secondary amine NH bend
1130-1190	Secondary amine CN stretch
1150-1210	Tertiary amine CN stretch
1250-1340	Aromatic primary amine CN stretch
1280-1350	Aromatic secondary amine CN stretch
1310-1360	Aromatic tertiary amine CN stretch

Table 4-9 Carbonyl compound functional groups

Wavenumber	Functional group associated to wavenumber
1550-1610	Carboxylate (Carboxylic acid salt)
1630-1680	Amide
1675-1690	Quinone or conjugated ketone
1700-1725	Carboxylic acid
1705-1724	Ketone
1726-1740	Aldehyde
1727-1750	Ester
1740-1760	Alkyl Carbonate
1770-1815	Acid (acyl) halide
1775-1820	Aryl Carbonate
1800-1850	Open chain acid anhydride
1860-2100	Transition metal carbonyls

Table 4-10 Nitrogen multiple double bond functional group

Wavenumber	Functional group associated to wavenumber
2240-2280	Aliphatic cyanide/nitrile
2220-2240	Aromatic cyanide/nitrile
2240-2260	Cynate (-OCN and C-OCN stretch)
2240-2276	Isocynate (-N=C=O asymmetric stretch)
2140-2175	Thiocyanate (-SCN)
1990-2150	Isothiocyanate (-NCS)
1590-1690	Open chain amino (-C=N-)
1575-1630	Open chain azo (-N=N-)

Table 4-11 Hetro-oxy compound functional groups

Wavenumber	Functional group associated to wavenumber
1540-1560	Aliphatic nitro compounds
1485-1555	Aromatic nitro compounds
1620-1640	Organic nitrates
1250-1350	Organic phosphates (P=O stretch)
990-1050	Aliphatic phosphates (P-O-C stretch)
850-995	Aromatic phosphates (P-O-C stretch)
1300-1335	Di-alkyl sulfones
1370-1420	Organic sulfates
1340-1365	Sulfonates
1075-1095	Organic siloxane or silicone (Si-O-Si)
1080-1110	Organic siloxane or silicone (Si-O-C)

Table 4-12 Thiols, Thio-substituted compounds and inorganic ions functional groups

Wavenumber	Functional group associated to wavenumber
2550-2600	Thiols (S-H) stretch
685-710	Thiol or Thio-ether, CH_2-S- (C-S stretch)
630-660	Thio-ethers, CH_3-S- (C-S stretch)
670-715	Aryl Thio-ethers, Φ-S (C-S stretch)
570-705	Disulfides (C-S stretch)
600-620	Disulfides (S-S stretch)
430-500	Aryl Disulfides (S-S stretch)
1410-1490	Carbonate ion
1080-1130	Sulfate ion
1350-1380	Nitrate ion
1000-1100	Phosphate ion
900-1100	Silicate ion

4.2 Analysis of Untreated/Raw Clay

Figure 4.1 shows the analysis of untreated clay. At lower wave numbers (from 600 to 1200 cm^{-1}) larger peaks formed which shows wide gap between the layers of clay structure but when the wave number is increased gaps become narrower and small peaks are formed. But between the range of 3600 to 3800 cm-1 peaks become wider little bit. At initial wave number range wider gaps are present due to presence of straight chain aliphatic hydrocarbons and alkyl halides. The spectrum range in raw clay analysis shows that there is lesser number of functional groups of unsaturated hydrocarbons and greater amount of straight chain hydrocarbons. Lesser amount of unsaturated hydrocarbons shows lower adsorption capacity.

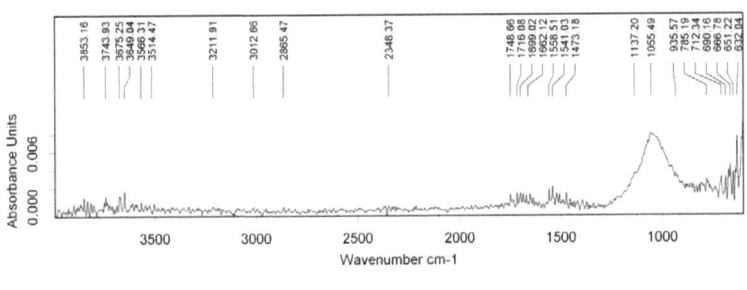

Figure 4-1 FT-IR analysis of untreated/raw clay

4.3 Clay analysis after treatment

Figure 4.2 shows the spectrum range of clay after treating with acetic acid. In this analysis peaks are reduced as compared to untreated clay. If acetic acid results are compared with phosphoric acid treatment (fig.4.3), more sharp peaks are produced but the result is same if compared with untreated clay analysis. Citric acid analysis (fig.4.4) shows same results but between the ranges (1600 to 2400 cm-1) more sharp peaks are formed which shows more unsaturated esters and saturated aliphatic hydrocarbons. Oxalic acid analysis (fig.4.5) shows the wide range of aromatics and phenolic groups more sharp peaks are formed between the ranges (2000 to 3800 cm-1). Oxalic acid analysis shows that more unsaturated organic functional groups are formed.

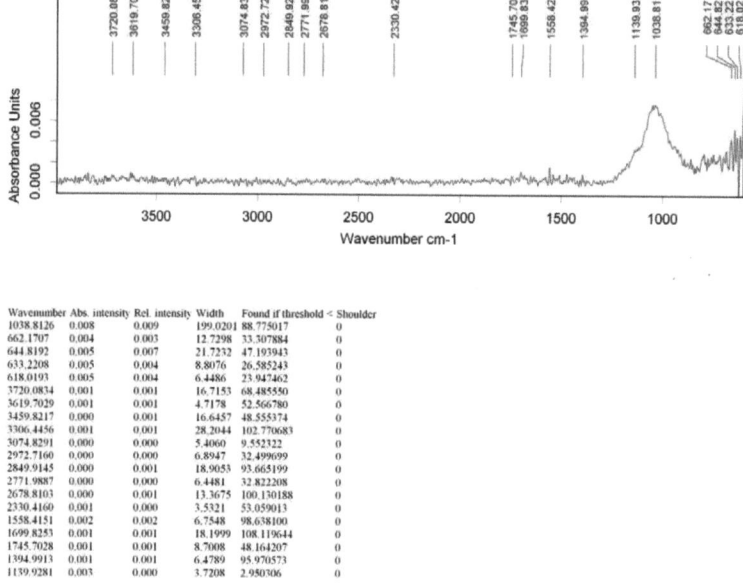

Wavenumber	Abs. intensity	Rel. intensity	Width	Found if threshold <	Shoulder
1038.8126	0.008	0.009	199.0201	88.775017	0
662.1707	0.004	0.003	12.7298	33.307884	0
644.8192	0.005	0.007	21.7232	47.193943	0
633.2208	0.005	0.004	8.8076	26.585243	0
618.0193	0.005	0.004	6.4486	23.947462	0
3720.0834	0.001	0.001	16.7153	68.485550	0
3619.7029	0.001	0.001	4.7178	52.566780	0
3459.8217	0.000	0.001	16.6457	48.555374	0
3306.4456	0.001	0.001	28.2044	102.770683	0
3074.8291	0.000	0.000	5.4060	9.552322	0
2972.7160	0.000	0.000	6.8947	32.499699	0
2849.9145	0.000	0.001	18.9053	93.665199	0
2771.9887	0.000	0.000	6.4481	32.822208	0
2678.8103	0.000	0.001	13.3675	100.130188	0
2330.4160	0.001	0.000	3.5321	53.059013	0
1558.4151	0.002	0.002	6.7548	98.638100	0
1699.8253	0.001	0.001	18.1999	108.119644	0
1745.7028	0.001	0.001	8.7008	48.164207	0
1394.9913	0.001	0.001	6.4789	95.970573	0
1139.9281	0.003	0.000	3.7208	2.950306	0

Figure 4-2 FT-IR analysis of acetic acid treated clay

Furthermore these clay analysis shows that the adsorption characteristics of clay has been modified very effectively due to the action of these acids. After treatment of the clay with these acids, more aromatic nitrile and phenolic groups are produced this shows the higher degree of unsaturation. Acetic acid treatment shows more sharp peaks with long wave lengths on the right side of the adsorption spectrum. These wavelength decreases as we go along the spectrum towards the left side. Sharp peaks on the right side of the spectrum shows that there are more straight chain aliphatic compounds especially long range of alkyl halides and their derivatives.

Similarly the smaller peaks on the left side of the spectrum shows the week tendency of aromatic and cyclic compounds which shows the lesser degree of saturation and hence week adsorption capacity. In phosphoric acid clay analysis same spectrum pattern is observed on right side. More sharp peaks having long wavelengths are produced showing strong tendency of alkyl halides, alkyl nitriles and their derivatives but these peaks becomes smaller and week as it approaches towards the left side and the relevant wavenumbers with small peaks shows the little tendency of aromatic and phenolic functional groups and their derivatives. But on the extreme left side of the spectrum these peaks become strong sharper having longer wavelengths compared to the previous behavior. So phosphoric acid treatment created little tendency of unsaturation in fuller earth and hence its absorption capacity is increased as compared to acetic acid treatment. If the analysis of citric acid is analyzed carefully then the same peak behavior is observed on the right side of the adsorption spectrum as observed in phosphoric acid and acetic acid treatment. But on the extreme left side peaks show very little tendency of exhibiting aromatic functional group derivatives, hence citric acid did not modify the adsorption properties of clay very well. According to the literature review citric acid is affective only on clay having specific properties. These properties of clay depend upon geographical locations as well. If oxalic acid analysis is closely observed there will be a clear difference of observation on the modification of clay properties. More sharp peaks are produced at certain height on the adsorption spectrum and peaks are sharper their number increases as we go from left to right in the adsorption spectrum. Although the wavelength of peaks are greater on the right side of adsorption spectrum but number of peaks formed are less if we compare it with left side. Oxalic acid treatment shows that adsorption capacity of clay is improved very well because it has more aromatic functional groups and their derivatives showing higher degree of unsaturation.

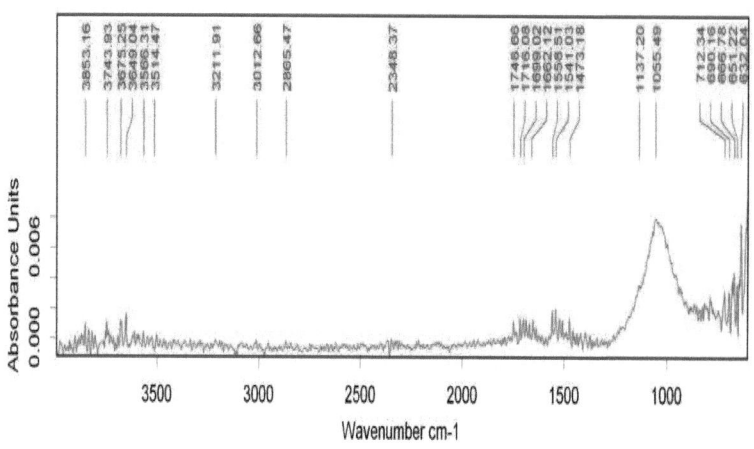

Wavenumber	Abs. intensity	Rel. intensity	Width	Found if threshold <	Shoulder
3649.0377	0.002	0.003	7.7245	26.565714	0
1541.0324	0.002	0.003	11.3585	26.858664	0
1055.4900	0.008	0.008	165.4525	73.787384	0
712.3435	0.003	0.003	13.0635	25.043816	0
690.1581	0.003	0.003	6.1187	20.570141	0
666.7806	0.005	0.004	15.3603	38.642677	0
651.2211	0.004	0.003	6.0675	25.540113	0
632.0407	0.008	0.005	5.5380	-15.227799	0
3853.1547	0.001	0.002	6.5354	101.899055	0
3743.9315	0.001	0.002	14.3313	86.844742	0
3675.2511	0.001	0.002	10.8534	69.509399	0
1748.6619	0.001	0.002	7.4371	81.118073	0
1716.0840	0.001	0.002	6.2801	101.278687	0
1699.0217	0.001	0.001	8.0564	79.410309	0
1662.1170	0.000	0.000	3.3822	9.097415	0
1558.5123	0.002	0.002	4.4109	61.927326	0
1473.1829	0.001	0.002	6.3182	71.579071	0
1137.2042	0.004	0.000	4.7575	3.493960	0
2348.3701	0.000	0.001	12.6800	74.786095	0
2865.4678	-0.000	0.001	15.2205	103.741264	0
3012.6590	-0.000	0.001	19.1677	99.195251	0

Figure 4-3 FT-IR analysis of phosphoric acid treated clay

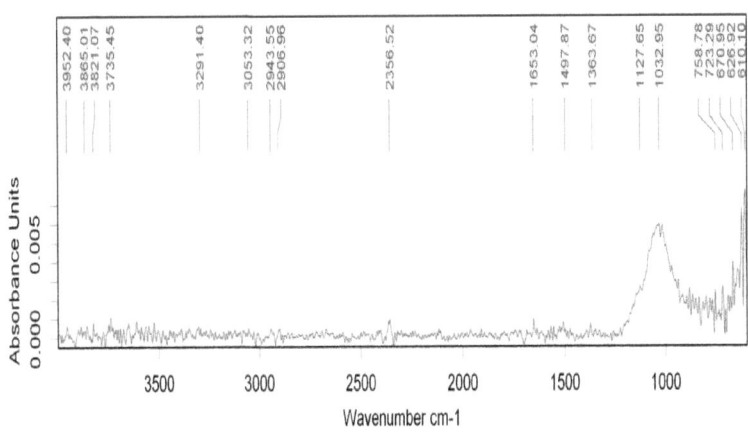

Wavenumber	Abs. intensity	Rel. intensity	Width	Found if threshold <	Shoulder
1032.9518	0.006	0.006	150.0828	72.064140	0
758.7843	0.002	0.002	6.0613	20.354040	0
723.2848	0.003	0.003	11.0173	31.536083	0
670.9527	0.004	0.003	6.5222	30.000929	0
626.9218	0.007	0.005	10.5934	66.040390	0
610.0948	0.008	0.004	9.0524	48.295856	0
3735.4544	0.001	0.002	27.8535	106.491417	0
3821.0679	0.001	0.001	8.0086	103.141090	0
3865.0081	0.000	0.000	7.5401	24.831602	0
3952.4003	0.001	0.001	14.0427	73.860123	0
2356.5208	0.001	0.001	24.7001	100.302368	0
1653.0405	0.001	0.001	11.9231	99.467010	0
1497.8701	0.000	0.000	4.2947	17.086750	0
1363.6717	0.000	0.000	3.5601	3.557847	0
1127.6533	0.003	0.000	5.0806	5.454212	0
3291.3947	0.001	0.000	4.2662	55.604580	0
2943.5481	0.000	0.001	27.7546	100.859650	0
2906.9551	0.000	0.000	13.5114	6.015716	0
3053.3231	0.001	0.001	32.4626	84.562920	0

Figure 4-4 FT-IR analysis of citric acid treated clay

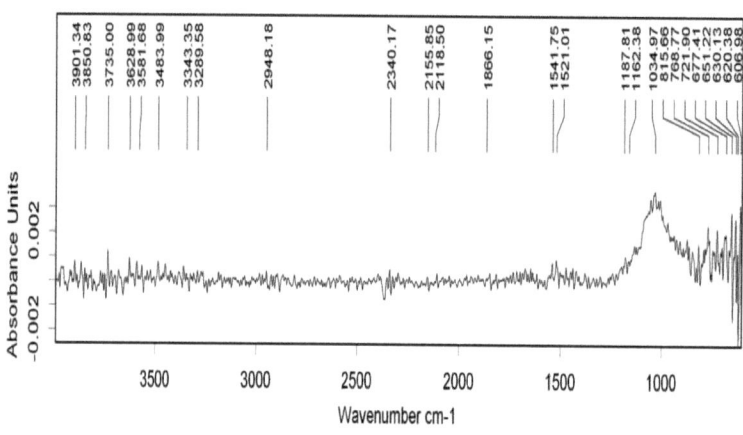

Wavenumber	Abs. intensity	Rel. intensity	Width	Found if threshold <	Shoulder
3901.3444	0.001	0.001	8.9923	20.511751	0
3850.8263	0.001	0.001	15.2273	20.350973	0
3735.0035	0.001	0.002	8.9063	31.996874	0
3628.9848	0.001	0.002	11.0440	24.367641	0
3483.9856	0.001	0.001	15.1916	20.759607	0
1034.9736	0.004	0.006	296.2088	71.665230	0
815.6620	0.002	0.002	7.1335	24.165712	0
768.7742	0.002	0.002	26.2409	36.280514	0
721.9000	0.002	0.002	8.9129	33.295330	0
677.4069	0.002	0.002	14.6640	29.306404	0
651.2187	0.003	0.005	17.8651	47.278172	0
630.1300	0.003	0.005	13.3729	67.985146	0
620.3769	0.001	0.003	4.8632	34.678017	0
606.9783	0.003	0.004	8.7562	51.151752	0
3343.3493	0.000	0.000	12.4558	34.189316	0
3289.5796	0.000	0.001	9.3078	81.733559	0
2948.1752	0.000	0.001	18.1699	106.473000	0
2340.1721	0.001	0.001	7.6700	84.911446	0
2118.5042	0.000	0.000	3.7912	11.683126	0
2155.8454	0.000	0.001	9.6778	61.567612	0
1866.1458	0.000	0.000	5.6762	10.074079	0

Figure 4-5: FT-IR analysis of oxalic acid treated clay

4.4 Conclusion & Recommendations

Organic acid treatment is the safest mode for surface activation of clay. Among four organic acids the best acid which showed good and satisfactory results is oxalic acid. Oxalic acid creates more unsaturated organic functional groups in bentonite clay (fuller's earth). Aromatic and phenolic functional groups are highly unsaturated functional groups which show the higher degree of adsorption capacity. Compared to other organic acids, Oxalic acid treatment modified the clay properties more effectively and enhanced the adsorptive properties of clay. So among these four organic acids oxalic acid is recommended acid for improving the adsorption properties of fuller's earth .Organic acid treated fuller's earth is also recommended for treatment of used lube oil. Because used lube oil can be regenerated by organic acid activated fuller's earth and it is very economical treatment method as compared to other conventional techniques.

References

[1] M. Rossi, M. Gianazza, C. Alamprese, and F. Stanga, "The role of bleaching clays and synthetic silica in palm oil physical refining," *Food Chemistry,* vol. 82, pp. 291-296, 2003.

[2] L. Riekert, "The efficiency of energy-utilization in chemical processes," *Chemical Engineering Science,* vol. 29, pp. 1613-1620, 1974.

[3] K. Beneke and G. Lagaly, "From fuller's earth to bleaching earth–a historical note," *ECGA Newsletter,* vol. 5, pp. 57-78, 2002.

[4] J. Madejová and P. Komadel, "Baseline studies of the clay minerals society source clays: infrared methods," *Clays and clay minerals,* vol. 49, pp. 410-432, 2001.

[5] F. Bergaya, B. K. Theng, and G. Lagaly, *Handbook of clay science* vol. 1: Elsevier, 2011.

[6] H. Babaki, A. Salem, and A. Jafarizad, "Kinetic model for the isothermal activation of bentonite by sulfuric acid," *Materials Chemistry and Physics,* vol. 108, pp. 263-268, 2008.

[7] M. Brigatti, E. Galan, and B. Theng, "Developments in clay science: handbook of clay science," *Structures and Mineralogy of Clay Minerals,* vol. 1, pp. 19-86, 2006.

[8] M. Vicente Rodriguez, J. DE D LOPEZ GONZALEZ, and M. Banares Munoz, "Acid activation of a spanish sepiolite: physicochemical characterization, free silica content and surface area of products obtained," *Clay minerals,* vol. 29, pp. 361-367, 1994.

[9] R. Mokaya and W. Jones, "Pillared clays and pillared acid-activated clays: a comparative-study of physical, acidic, and catalytic properties," *Journal of Catalysis,* vol. 153, pp. 76-85, 1995.

[10] G. Habashy, A. Gadalla, T. Ghazi, W. Mourad, and S. Nashed, "Characterization of some Egyptian clays to be used as bleaching agents," *Surface Technology,* vol. 15, pp. 313-322, 1982.

[11] M. Didi, B. Makhoukhi, A. Azzouz, and D. Villemin, "Colza oil bleaching through optimized acid activation of bentonite. A comparative study," *Applied Clay Science,* vol. 42, pp. 336-344, 2009.

[12] A. Al-Zahrani, S. Al-Shahrani, and Y. Al-Tawil, "Study on the activation of Saudi natural bentonite, part I: investigation of the conditions that give best results and kinetics of the sulfuric acid activation process," *J. King Saud. Univ,* vol. 13, pp. 57-72, 2001.

[13] F. Hussin, M. K. Aroua, and W. M. A. W. Daud, "Textural characteristics, surface chemistry and activation of bleaching earth: A review," *Chemical Engineering Journal,* vol. 170, pp. 90-106, 2011.

[14] J. Coates, "Interpretation of infrared spectra, a practical approach," *Encyclopedia of analytical chemistry,* 2000.

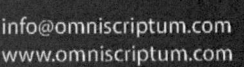

Printed by Books on Demand GmbH, Norderstedt / Germany